small worlds

A TIDAL POOL

Philip Steele

CRABTREE
Publishing Company

Crabtree Publishing Company

350 Fifth Avenue
Suite 3308
New York, NY 10118

360 York Road, R.R.4
Niagara-on-the-Lake
Ontario LOS 1J0

Co-ordinating editor: Ellen Rodger
Commissioning editor: Anne O'Daly
Editor: Clare Oliver
Designer: Joan Curtis
Picture researcher: Christine Lalla
Consultants: Staff of the Natural History Museum, London
and David T. Brown PhD

Illustrator: Peter Bull
Photographs: Mark N Boulton/Bruce Coleman Limited p 10*b*; Jeff Foott/Bruce Coleman Limited pp
4*t*, 13; Charles & Sandra Hood/Bruce Coleman Limited pp 6*b*, 25; Harald Lange/Bruce Coleman
Limited pp 6*t*, 16; Gordon Langsbury/Bruce Coleman Limited pp 8*m*, 27*b*; Dr Eckart Pott/Bruce
Coleman Limited pp 15, 26*t*; Andrew J Purcell/Bruce Coleman Limited p 21; Frieder Sauer/Bruce
Coleman Limited pp 11, 14; Kim Taylor/Bruce Coleman pp 10*t*, 20, 26*b*, 27*t*; Bruce Coleman p 12;
Corbis Images, front and back cover, pp 3, 4*m*, 5, 17*b*, 19*t*, 23, 30, 31; Joe Devenney/Imagebank p 29;
Derek Redfern/Imagebank p 28*t*; Daniel Heuclin/NHPA, front cover, pp 1, 19*b*; G J Cambridge/NHPA
p 8*t*; N A Callow/NHPA p 17*t*; Laurie Campbell/NHPA p 22; NHPA p 28*b*; Harry Smith Horticultural
Collection pp 7, 9, 24.

Created and produced by
Brown Partworks Ltd

First edition 1999
10 9 8 7 6 5 4 3 2 1

CATALOGING-IN-PUBLICATION DATA

Steele, Philip, 1948-
    A tidal pool / Philip Steele. — 1st ed.
    p. cm. — (Small worlds)
Includes index.
    SUMMARY: Describes the different plants and animals that live in tidal pools at low-, mid-, and
high-tides.
    ISBN 0-7787-0135-2 (rlb)
    ISBN 0-7787-0149-2 (pbk.)
    1. Tide pool ecology— Juvenile literature. 2. Tide pool animals—Habitat—Juvenile literature. 3. Tide
pools—Juvenile literature. [1. Tide pool ecology. 2. Tide pool animals. 3. Ecology.] I. Title. II. Series:
Small worlds.
    QH541.5.S35 S75 1999
    577.69'9—dc21

LC 98-51708
CIP
AC

Printed in Singapore

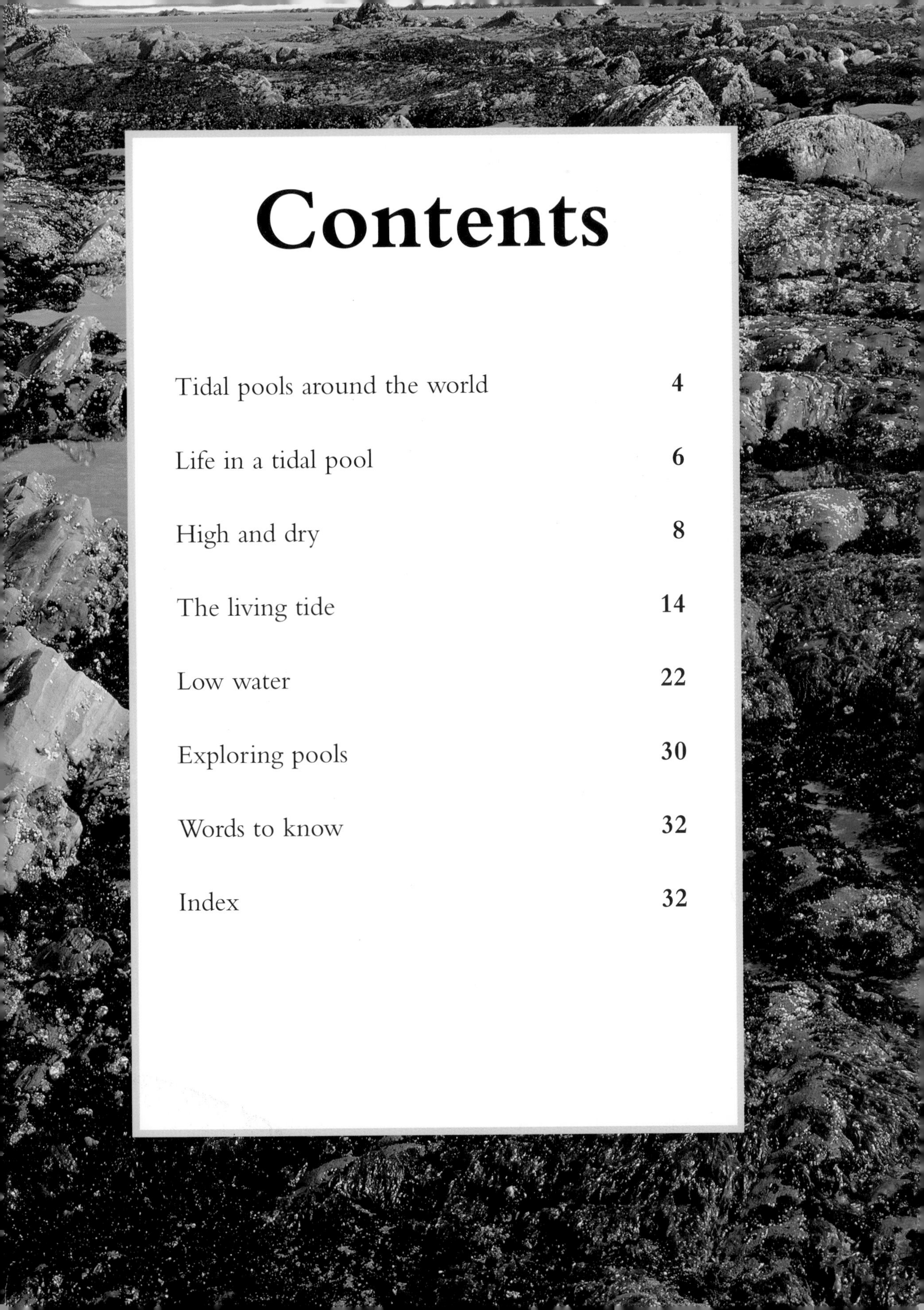

Contents

Tidal pools around the world   4

Life in a tidal pool   6

High and dry   8

The living tide   14

Low water   22

Exploring pools   30

Words to know   32

Index   32

Tidal pools around the world

Welcome to the place where the land meets the ocean. Swirling tides crash to shore, then roll back out to sea again.

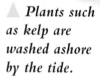

▲ *Plants such as kelp are washed ashore by the tide.*

▶ *As the Earth, Sun, and Moon travel through space, they set up a force called gravity, which tugs at the oceans. Gravity makes the water rise and fall, causing tides.*

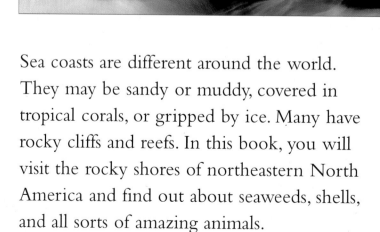

Sea coasts are different around the world. They may be sandy or muddy, covered in tropical corals, or gripped by ice. Many have rocky cliffs and reefs. In this book, you will visit the rocky shores of northeastern North America and find out about seaweeds, shells, and all sorts of amazing animals.

▶ *No two tidal pools are the same. Each one has its own environment.*

Life in a tidal pool

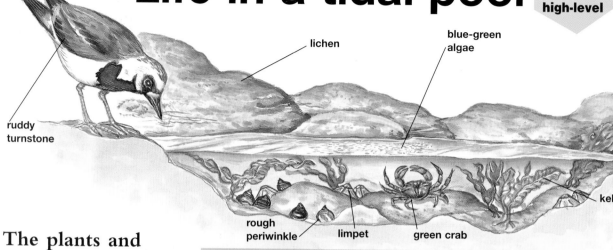

- ruddy turnstone
- lichen
- blue-green algae
- rough periwinkle
- limpet
- green crab
- kelp

The plants and animals that live in tidal pools are survivors.

They must withstand the pounding surf and risk being left high and dry on the rocks. Conditions, such as saltiness, temperature, or the amount of oxygen in the water, can change in an instant.

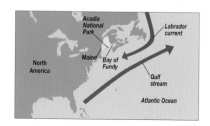

Waters off the northeastern coast of North America are warmed by the Gulf stream.

The living tide

Mid-level tidal pools make an ideal home for small fish and sea anemones. Limpets cling to the rocks, and sand shrimps dart among the seaweeds. Seawater, full of foaming air bubbles, floods into mid-level tidal pools twice a day.

Low water

Low-level tidal pools uncover reefs and ledges that are rarely seen. These are often blanketed in brown kelp. Life in low-level tidal pools is much the same as that in the open ocean. There may be lobsters and fish, such as the lumpfish. Noisy gulls squabble over and scoop up a meal of rock crabs.

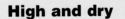

High and dry

Tidal pools on high ledges, or
at the top of a beach, fill up
only when there are storms, floods, or very high tides.
Some get water from flying surf. These tidal pools lie in
the "splash" zone. Survival is at its toughest here. Pools
may dry out or fill with rainwater and lose their saltiness.

herring gull

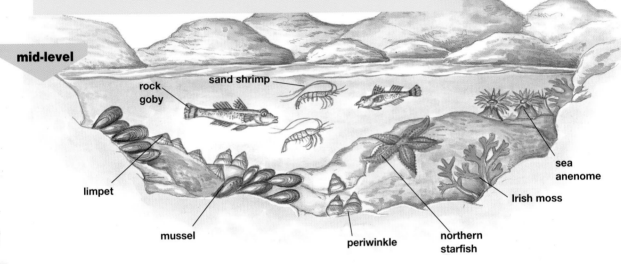

mid-level

rock
goby

sand shrimp

sea
anenome

Irish moss

limpet

mussel

periwinkle

northern
starfish

striped pink
shrimp

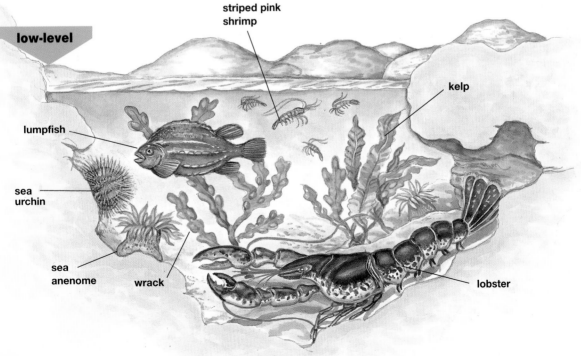

low-level

kelp

lumpfish

sea
urchin

sea
anenome

wrack

lobster

High and dry

▲ *The rock louse lives just above the high-tide level. It comes out at night to eat seaweed.*

Creatures of high-level tidal pools must put up with long periods of drying out, interrupted by sudden floods or storms.

▶ *The ruddy turnstone, a winter visitor to rocky shores, turns over stones and seaweed in search of tasty morsels.*

In a high-level pool, hours of wind and sunshine may make the seawater **evaporate** or turn into vapor. When it evaporates, the salt it contains stays behind. As the water level drops, the pool becomes saltier and saltier. It is very different on rainy days. The pool fills up with fresh water and becomes less salty.

▶ *Rocks may be covered in crusty **lichens**. Lichens are made up of two kinds of plants: **algae** and fungi.*

Because life at the upper levels is tough, these pools are home to fewer types of animals than the lower ones. Even so, the tiny creatures that do live there are often present in very large numbers. Insects called marine springtails, which have waterproof bodies just one sixteenth of an inch (0.16 cm) long, swarm across many pools.

▲ *The water in high-level pools is calmer than in the surf-filled lower pools. Often, a film builds up on the surface of the water. The springtails shown here are light enough to walk on this film without sinking.*

Armor-plated barnacles

Barnacles can live above or below the high-tide level. These little creatures are cemented firmly to the rocks and are covered in a chalky shell made up of six, tightly fitting plates. The plates lock in the moisture and also protect the creatures inside from the crashing surf and hungry predators. When the tide comes in, the barnacle opens its plates to let out six pairs of waving legs. With these, it fans scraps of food from the water into its mouth.

▶ *Safe inside its armored shell, the barnacle's body is soft and slimy.*

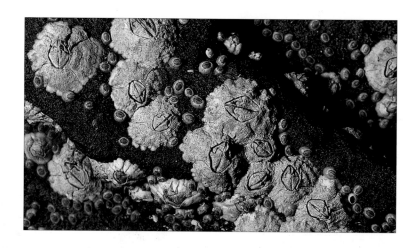

Hiding away

How do creatures survive when the high-level pools are drying out? Some of them squeeze into deep crevices in the rock, where it is always damp and dark. Some hide in clumps of wet seaweed. Many animals come out only at night, when it is cool and moist.

Finding the right spot

The rough periwinkle is a sea snail, one of the group of animals called **molluscs**. It is able to sense exactly how high it is on the shore. When conditions become too dry, it pulls back into its shell and seals off the opening, just as land snails do during dry weather.

FANTASTIC FACTS

● Tides at Acadia National Park, Maine, vary between heights of nine and 15 feet (2.7 and 4.6 m).

● Periwinkles have gills, so they can take oxygen from the water. They trap extra water and oxygen inside their shells so that they can survive outside the water for long periods.

These rough periwinkles are clustered together on their favorite food—seaweed.

The green crab has ten legs. The front two form powerful pincers for crushing and tearing food.

Hunters of the upper shore

Crabs belong to a group of animals called **crustaceans**. Rock lice, barnacles, lobsters, and shrimps are also crustaceans. By day, green crabs hide away under the boulders and weeds. They come out by night in search of food, such as molluscs or animal remains.

Green crabs normally live on their own. They are not very big, growing only to about one and a half inches (3.8 cm) long. Adult males are dark green to black in color, but females and young crabs come in all sorts of patterns and colors.

The green crab can hunt anywhere on the shore but likes the upper tidal pools best. Unlike its swimming, deep-water relatives, it can scuttle over dry land. It can cope with lower salt levels than other crabs can.

FANTASTIC FACTS

● The highest tides happen when the Sun, Moon, and Earth all move into a straight line.

● Life on Earth began in the sea. Over 350 million years ago, some upper shore creatures moved inland from their pools and became the first ever land animals.

On the fringes

All sorts of algae live on high rocky shores.
Algae are simple plants that range in size from
tiny threads and fronds to large seaweeds.

Tufts of green algae can be seen waving in
the water, while a tangle of tough brown-orange
plants called wracks are often found on the
fringe of the high-tide level or are washed up
by storms. The wracks provide food for small
animals and a place to hide for many more.

Algae control the living conditions in many
pools. During the day, green algae use the
energy from the sunlight to make the food
plants need to survive. In doing this, they give
out life-giving oxygen. However, during the
night the oxygen in the pool gets used up, and
the pool fills with **carbon dioxide**. This is a gas
that can poison many living things, but not the
tough survivors of the upper-shore tidal pools.

▼ *Delicate green
algae grow best at
the bottom of pools,
but tough wracks
can survive in wind
and sunshine.*

The living tide

Plants and animals that are located in pools along the middle shore are less likely to dry out. The tide sweeps over them twice a day.

The northern starfish has five legs, called rays. *If one leg breaks off, the starfish simply grows another one.*

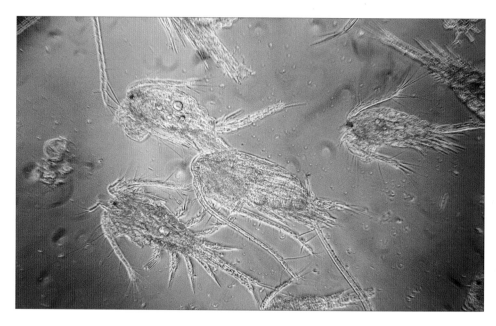

Plankton float freely with the currents and tides. Many tidal pool creatures feed on plankton.

Tidal pools that are regularly covered by clean, salty water are home to an amazing range of creatures and are visited by many more. Each fresh tide brings in food, in the form of **plankton**. This is a floating soup made up of microscopic plants and animals, as well as the eggs and tiny **larvae** that are the early stages of many ocean life-forms. Some larvae stay in the tidal pool until they grow into their adult forms.

Mussels feed by filtering tiny bits of food from the water. A mussel can pump up to 15 gallons (57 l) of seawater through its body in a single day.

Limpet clamp-down

In and around the mid-level pools, the rocks may be covered in small, chalky, cone-shaped shells. These are called limpets. They clamp themselves onto the rocks so tightly that they cannot be lifted off, even by the strongest waves. Sometimes you can even see circles where a limpet shell has ground its way into the solid rock.

Limpets look like large barnacles, but they are really molluscs. Underneath that shell is a large, powerful "foot." The limpet uses this foot to creep about on the rock surface as it eats the algae there. It uses its tongue, called a radula, to scrape its food off the rocks.

Browns, reds, and purples

Tough knotted wracks line the upper rocks of mid-shore tidal pools. Some wracks have bladders in their fronds (leaf-like shoots). The bladders contain air bubbles, which help the wracks to ride the surf.

Red and purple seaweeds grow further down in the tidal pool. The seaweed thrives underwater and cannot survive the bright sunlight and dry spells of the high-level pools.

Mussel power

Did you know that mussels have beards? The beards are the brown threads that firmly anchor the mussels' blue-black shells to the rocks. Mussels hang in great clusters around mid-level and low-level tidal pools. They are bivalve molluscs, which means that their shells have twin halves. The curved shape of the mussel shell is perfectly designed for slipping through crashing breaker waves.

▲ *Irish moss is not a moss at all! It is one of the seaweeds found on middle and lower level rocks. It contains a chemical that is harvested and used to thicken foods such as ice cream.*

Beautiful stingers

Sea anemones are named after flowers. Long ago, people believed that they must be plants, with their colorful, waving leaf-like fronds. In fact, they are animals related to jellyfish, and their fronds are deadly **tentacles** lined with stingers. Sea anemones live underwater, although some can survive on open rocks by pulling in their tentacles between tides.

▼ *Sea anemones feed on striped pink shrimps, marine worms, and other small tidal pool residents.*

Everybody is somebody's dinner

Tidal pools are small worlds or environments where the plants and animals all depend upon each other. A herring gull might gobble up a starfish, which has just eaten a mussel, which has just eaten plankton. Each organism or life-form is a link in a food chain. Connect up all the different tidal pool food chains, and you have what is called a food web—a network of survival.

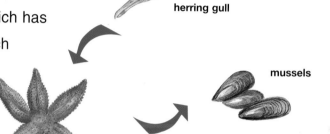

herring gull

mussels

starfish

Stars and spikes

Many types of starfish can be found in mid- and low-level tidal pools. Of all the creatures to be found here, these are some of the most amazing.

To eat a mussel, a starfish starts off by gripping the shell with the suckers on its feet. It forces open a narrow gap. Then it turns its own stomach inside out and pushes this out through its mouth and into the gap in the mussel's shell. The stomach starts to digest the mussel. Then, when the meal is over, the starfish swallows its own stomach again!

FANTASTIC FACTS

● The brittle star is a relative of the starfishes. If one of its long, spindly rays is grabbed by a hungry animal, it simply casts it off and moves away on the others.

● Most starfish have five rays, but some North American starfish have up to 50!

The starfish has some odd relatives. Brittle stars have long, thin rays, sea urchins bristle with sharp spines, and the sea cucumber is long and spiky.

Crusty crustaceans

The crustaceans that live in mid-level pools are smaller than the giants such as lobsters that live offshore. Sand shrimps and striped pink shrimps dart in and out of the weeds and boulders on the bottom.

Most crustaceans have hard shells. As their bodies grow bigger, they shed the shells and grow new ones. The hermit crab just steals the shell of a mollusc and squeezes into that. When the hermit crab outgrows its shell, it looks for a bigger one!

▲ The green sea urchin's spines are just over half an inch (1.3 cm) long.

▼ A hermit crab pushes its soft rear end into an empty shell. The front of the body is armor-plated against an attack.

Fooling the enemy

Many of the creatures that live in tidal pools are very difficult to see. That is because they are so well **camouflaged**. Their coloring blends in with the stones and seaweeds in which they live, and this protects the animals from their enemies. Many molluscs and crabs have seaweeds and sponges growing from their shells. This makes them look like rocks and pebbles. Some shrimps, like the one below, are see-through and seem invisible in the water.

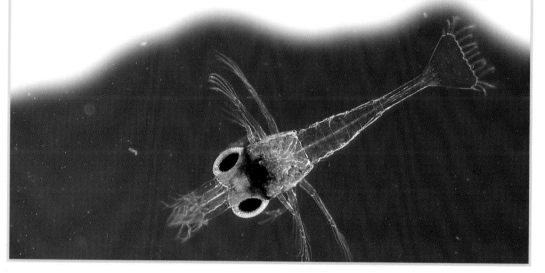

▶ *Rock gobies live at the bottom of tidal pools. Even when they rest, the fish are on the lookout for predators.*

Small fry

Small fish may be found among the weeds and stones of many mid-level tidal pools. Some of these may be stranded by high tides, but others live in the pools all the time. Rock gobies and blennies are typical mid-level dwellers.

Most types of goby are about three inches (7.6 cm) long. Two of their fins join to form a sucker. This helps them to grip the crevices in the rock, where they hide.

Living together

Some gobies share the cracks and crevices where they live with other little creatures, such as sand shrimps. The shrimps look after their shared home while the goby acts as a guard at the entrance. A lot of tidal pool creatures help each other out in one way or another. Sea anemones will settle on the shell of a hermit crab. The anemones' tentacles protect the hermit from attack. In return, the anemones feed on little pieces of the crab's dinner.

FANTASTIC FACTS

● The goby's eyes are placed high on the sides of its head so that it can see upward and check for danger.

● There are about 1,500 different types of goby.

● Sand gobies get their name from their habit of burying themselves under the sand to hide.

Low water

At the ocean's edge there are jagged reefs and rock ledges, which are uncovered only by the lowest tides.

▲ Herring gulls scan the lower shore for food. They pick up sea urchins, then drop them onto the rocks below to smash the shells open.

▶ The bread-crumb sponge looks like a piece of seaweed but is an animal. It filters food from the water, which it pumps out through mini "volcanoes."

The lowest tidal pools rarely dry out. Even at low tide, strong winds may drive waves over some of them. The rocks near the seashore take the full force of ocean breaker waves and currents. These tidal pools are the stormy frontier between the land and the sea. They are a danger zone for people and ships but a safe haven for marine life, such as sponges and crabs. Herring gulls often feed on the fish and crabs that live in these pools.

▶ Kelp can be seen most clearly at low tide. It forms long, broad ribbons and rubbery, brown belts.

▶ *An underwater forest of kelp is home to all sorts of creatures, just like a forest on dry land.*

The kelp forest

As the sea rolls back over the lower shore, large heaps of brown seaweed, called kelp, are draped over the rocks. When the tide comes back in, the kelp rises up to form enormous underwater forests. Kelp needs the energy from sunlight to make food, so it grows best in clear water.

It grows near big rocks with gentle waves and currents, rather than open surf.

The kelp forest is home to countless marine worms, crabs, jelly-like sea squirts, and fish such as wrasse. Red seaweeds and sponges live on the kelp itself. Worms, molluscs, and sea urchins all feed on the kelp's fronds and on its stalk, which is called a stipe.

FANTASTIC FACTS

● Over 300 different types of animals may live in and on a single kelp plant.

● Some tube worms spend their whole life inside the same kelp root.

● Atlantic kelps are dwarfs compared with the Pacific giant kelp, which grows to 200 feet (61 m).

Anchor in a storm

Just listen to the tide gurgling, slapping, sucking, and crashing along a rocky shore. The currents can be fierce. Many plants and animals have only survived by developing strong anchors. Seaweeds such as kelps and wracks grip the rocks tightly with a kind of root called a holdfast.

Sea urchins have suckers on their feet. These are so strong that the animals can walk up a sheer rock face. The lumpfish, shown below, has a powerful sucker on its belly. Some marine worms find shelter from the storm by boring a tunnel for themselves out of solid rock!

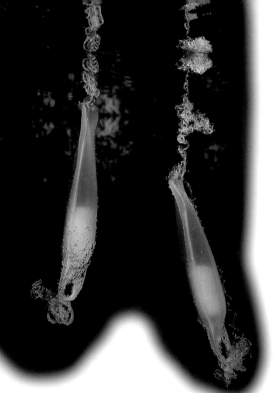

You may find the large egg cases of the nurse shark tangled in the kelp of the lower shore. The nurse shark is a type of dogfish. It breeds in shallow waters, and the egg cases float ashore.

The breeding grounds

Many animals use low-level tidal pools to breed. All sorts of fish and molluscs lay their eggs in these pools. The eggs of most small sea creatures are only made fertile when the male releases his **sperm** into the water around them. When the larvae hatch out, the tides carry them away as plankton.

Why is the lower shore best for breeding? Remember, the top end of the shore is not flushed out so often by the tides. It may be far from the open sea. Most eggs need to be kept at a steady temperature. They cannot survive the extreme changes of the upper pools.

Not all tidal pool dwellers produce eggs and larvae. Some sea anemones simply bud off and produce miniature versions of themselves.

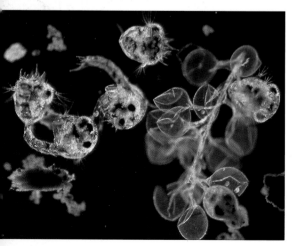

Safety in numbers

Most eggs and larvae are eaten by sea gulls or crows along the shoreline or by crabs, molluscs, and fish. To make up for this, some sea creatures guard their

Baby rock crabs are too tiny to see when they hatch. As adults, they grow to more than six inches (15.2 cm) across the shell.

This female striped pink shrimp carries about 2,000 eggs, which are stuck firmly to her body for safety.

eggs very closely. Many lay large numbers of eggs, just to make sure that enough will survive. A female rock crab lays about three million eggs in its lifetime. These hatch out and the young grow in low-level pools before moving to the open seabed.

Eider ducks dive for crabs and mussels, which they pull off the underwater rocks.

Jellyfish are related to sea anemones. They drift on the surface of the sea. They are often stranded in low pools and on beaches, but they cannot survive out of the water.

Out to the ocean

Just as the upper pools are on the edge of dry land, the lower pools border another environment, too—the open ocean. Creatures that normally live offshore often find their way into lower pools and gullies. You might see a rock crab, a lobster, a washed-up jellyfish, or the remains of a large fish in an upper pool.

The lower shore may be the closest environment to that of the open sea, but there are great differences, too. As the seabed descends from the shore, the waters become dimmer.

The lobster larva, less than a quarter of an inch (6 mm) long, drifts among the plankton. It later sinks to the rocks of the seabed. The Maine coast is famous for its lobsters.

At a depth of 25 feet (7.6 m), nearly all the surface sunlight is screened out. The kelp forests cannot survive in this gloom. The world of the shoreline has given way to that of the vast, mysterious ocean.

Changing coastlines

Tidal pools may look as if they have been there forever, but they are part of an endless process of change. The waves and tides eat away at the rocks, wearing them down into reefs and pools. Rocks and shells are ground down into grit and sand. Over time, these build up again and are squeezed into new rock formations. The water levels of the world's oceans constantly rise and fall. As coastlines sink and rise again, new homes are made for sea creatures.

Exploring pools

If you live by the sea, or go there on holiday, you can have a closer look at the amazing world of tidal pools.

Always handle tidal pool creatures very gently. Tip them back into their tidal pool home when you have finished having a close look at them.

Next time you visit the seashore, explore the tidal pools. You may want to collect empty shells or seaweeds, but it is much more exciting to watch the creatures and plants in the pool. A face mask and a snorkel will help you to see things more clearly. A net and a bucket are useful, too.

Sea urchin skeletons are popular with some beach tourists, but they are much more beautiful when you see them alive in a tidal pool.

Never take away or harm living creatures. Remember to replace any boulders you overturn. Tidal pool animals that make their homes on the underside of a rock may die if they are left out in the air.

Over 1,215 types of mollusc are in danger around the world today. More than 286 types have died out, or become extinct, in the last 150 years. Some have been collected as tourist souvenirs. Others have been killed off by pollution, the poisoning of our seas and shores. So take care not to do any more damage to these magical, miniature worlds!

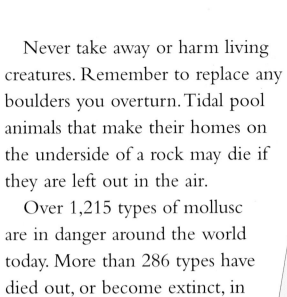

WARNING ALL POOL EXPLORERS!
Rocky shores are dangerous places. The safest pools to explore are those of the upper and middle shore. Always tell an adult where you are going.
• Beware of slippery, weed-covered rocks.
• Beware of crumbling cliffs.
• Beware of climbing high rocks.
• Don't get cut off by the tide.
• Watch out for sea urchin spines and jellyfish stings.

Words to know

algae Very simple plants.
camouflage The way an animal's coloring hides it.
carbon dioxide A gas taken in by plants.
crustacean One type of animal with its skeleton outside its body, such as a crab or shrimp.
evaporate To dry out.
larva An early stage in the life of a small creature.
lichen A mixture of algae and fungi that grows on rocks.

mollusc A soft-bodied animal, such as a clam or squid, which may or may not have a shell.
oxygen A life-giving gas found in air and water.
plankton Tiny plants and animals.
ray The arm of a starfish.
sperm A reproductive cell given out by a male, which fertilizes a female's eggs.
tentacle The arm of a sea anenome.

Index

algae 6, 8, 13, 18, 32
barnacle 10, 12
eider duck 27
green crab 6, 12
herring gull 6, 7, 18, 22
hermit crab 19, 21
Irish moss 7, 17
jellyfish 17, 28
kelp 4, 7, 22, 24, 25
lichen 6, 8, 32

limpet 6, 7, 16, 18
lobster 6, 7, 12, 28
lumpfish 6, 7, 25
marine springtail 10
mussel 7, 14, 17, 18
nurse shark 26
periwinkle 6, 11
plankton 14
rock crab 6, 26, 27
rock goby 7, 20, 21
rock louse 8, 12

ruddy turnstone 6, 8
sea anenome 6, 7, 17, 21, 26
sea urchin 7, 19, 22, 25, 31
seaweed 4, 7, 11, 13, 16, 17, 20, 24, 25
shrimp 6, 7, 12, 17, 19, 20, 21, 27
sponge 20, 22, 24
starfish 7, 14, 18